MULTIPLICATION

FACTS

COLOURING BOOK

1-12

THE EASY WAY TO LEARN THE TIMES TABLES

Multiplication Facts Colouring Book 1-12: The Easy Way to Learn the Times Tables

ISBN 978-1-897384-80-0

Magdalene Press, 2016

MULTIPLICATION

FACTS

COLOURING BOOK

THE EASY WAY TO
LEARN THE
TIMES TABLES

1-12

1X

1X1=1

1X2=2

1X3=3

$1 \times 4 = 4$

$1 \times 5 = 5$

$1 \times 6 = 6$

$1 \times 7 = 7$

$1 \times 8 = 8$

$1 \times 9 = 9$

$1 \times 10 = 10$

$1 \times 11 = 11$

1X12=12

2X

2X1=2

2X2=4

2X3=6

$$2 \times 4 = 8$$

$$2 \times 5 = 10$$

$$2 \times 6 = 12$$

$$2 \times 7 = 14$$

$2 \times 8 = 16$

$2 \times 9 = 18$

$2 \times 10 = 20$

$2 \times 11 = 22$

2X12=24

3X

$3 \times 1 = 3$

$3 \times 2 = 6$

$3 \times 3 = 9$

3 X 4 = 12

3 X 5 = 15

3 X 6 = 18

3 X 7 = 21

$$3 \times 8 = 24$$

$$3 \times 9 = 27$$

$$3 \times 10 = 30$$

$$3 \times 11 = 33$$

3X12=36

4X

4X1 = 4

4X2 = 8

4X3 = 12

$$4 \times 4 = 16$$

$$4 \times 5 = 20$$

$$4 \times 6 = 24$$

$$4 \times 7 = 28$$

4 X 8 = 32

4 X 9 = 36

4 X 10 = 40

4 X 11 = 44

4X12=48

5X

5X1=5

5X2=10

5X3=15

5X4=20

5X5=25

5X6=30

5X7=35

$$5 \times 8 = 40$$

$$5 \times 9 = 45$$

$$5 \times 10 = 50$$

$$5 \times 11 = 55$$

5X12=60

6x

6x1=6

6x2=12

6x3=18

6X4=24

6X5=30

6X6=36

6X7=42

6 X 8 = 48

6 X 9 = 54

6 X 10 = 60

6 X 11 = 66

7X

7 X 1 = 7

7 X 2 = 14

7 X 3 = 21

7 X 4 = 28

7 X 5 = 35

7 X 6 = 42

7 X 7 = 49

7×8=56

7×9=63

7×10=70

7×11=77

7x12=84

8X

8X1 = 8

8X2 = 16

8X3 = 24

$$8 \times 4 = 32$$

$$8 \times 5 = 40$$

$$8 \times 6 = 48$$

$$8 \times 7 = 56$$

$$8 \times 8 = 64$$

$$8 \times 9 = 72$$

$$8 \times 10 - 80$$

$$8 \times 11 = 88$$

9X

9X1=9

9X2=18

9X3=27

$$9 \times 4 = 36$$

$$9 \times 5 = 45$$

$$9 \times 6 = 54$$

$$9 \times 7 = 63$$

$9 \times 8 = 72$

$9 \times 9 = 81$

$9 \times 10 = 90$

$9 \times 11 = 99$

9X12=108

10X

$$10 \times 1 = 10$$

$$10 \times 2 = 20$$

$$10 \times 3 = 30$$

$$10 \times 4 = 40$$

$$10 \times 5 = 50$$

$$10 \times 6 = 60$$

$$10 \times 7 = 70$$

$$10 \times 8 = 80$$

$$10 \times 9 = 90$$

$$10 \times 10 = 100$$

$$10 \times 11 = 110$$

10X12=120

11X

11X1 = 11

11X2 = 22

11X3 = 33

11X4=44

11X5=55

11X6=66

11X7=77

$$11 \times 8 = 88$$

$$11 \times 9 = 99$$

$$11 \times 10 = 110$$

$$11 \times 11 = 121$$

11X12=132

12X

12X1=12

12X2=24

12X3=36

$$12 \times 4 = 48$$

$$12 \times 5 = 60$$

$$12 \times 6 = 72$$

$$12 \times 7 = 84$$

12X8=96

12X9=108

12X10=120

12X11=132

12X12=144

Congratulations!

You did it!

www.ingramcontent.com/pod-product-compliance
Lightning Source LLC
Chambersburg PA
CBHW080722220326
41520CB00056B/7376